# Node.js Web app development

I0468388

## Create Your Own Web Apps with Node.js Platform

2nd edition

**By Daniel Green**

## Disclaimer

While all attempts have been made to verify the information provided in this book, the author doesn't assume any responsibility for errors, omissions, or contrary interpretations of the subject matter contained within. **The information provided in this book is for educational and entertainment purposes only. The reader is responsible for his or her own actions and the author does not accept any responsibilities for any liabilities or damages, real or perceived, resulting from the use of this information.**

The trademarks that are used are without any consent, and the publication of the trademark is without permission or backing by the trademark owner. All trademarks and brands within this book are for clarifying purposes only and are the owned by the owners themselves, not affiliated with this document.

# Contents

# Book Description

This book is an exploration of Node js in detail. It begins by explaining the origin of the platform, its purpose, and how it is used. You will be introduced to those behind the development of the platform, and those responsible for its hosting. The second part of the book is a guide on how you can set up the environment ready for programming.

The book will guide you on how to choose the best editor, depending on the kind of operating system that you are using. The next step is a guide on how to install the Node js package on your system. Installation in Linux, Mac OS X, and Windows is explored, providing detailed instructions for doing so. . A Node js application is made up of different parts.

After reading this book, you will know how to create these parts, including the server. The REPL terminal, which is a very powerful and important component of Node js, is examined in detail. You will learn how to run mathematical expressions on REPL, as well as how to write complex programs on it.

The book also provides a guide on how you can access REPL on your system. Callbacks, which are kinds of functions in Node js, are explored in detail, and thus, you will know how to use these in your programs. Event loops and Event emitters are then discussed. The book will also guide you on how to perform various operations on streams and files in Node js.

The following topics are explored:

- Definition

- Environmental Setup

- Creating our First Application

- The REPL Terminal

- Callbacks in Node.js

- Event Loop

- Event Emitter

- Buffers

- Streams in Node.js

- The File System

# Introduction

Node js is purely based on Javascript. It is well known for its strong support for real time and data intensive applications. This is why the platform is loved by many. The good thing about the platform is that it can be supported on the various operating systems that are used by developers. These include Linux, Mac OS X, and Windows. There are also many other operating systems which support this. The platform is broadly used by developers for the development of networking applications. The reason is because it uses a non-blocking and event-driven I/O, which is the reason behind its ease of use and being fast.

# Chapter 1:
# Definition

This is a platform which was built on the Javascript runtime for Chrome. It was developed with the aim of making the process of developing network applications which are scalable easy and fast. The platform is efficient and lightweight, due to the fact that it uses a non-blocking and event-driven I/O. The platform is also suitable for real-time and data-intensive applications which run on devices which are distributed.

The platform is open-source, meaning that you can download and use it for free on your system. Developers use this platform to create server-side and networking applications. Applications written in this platform exhibit a wide range of flexibility, since they can be run on a variety of operating systems such as Windows, OS X, FreeBSD, Linux, NonStop, and others. The Node.js Foundation is responsible for hosting the work of this platform.

With this platform, the scalability and the throughput of an application is greatly optimized. The reason for this is due to the fact that it provides a non-blocking API, and an event-driven architecture. These features are important as far as real-time web applications are concerned. In Node.js, the Google V8 Javascript engine is used for the purpose of execution of the code.

Javascript is used for the writing of most of the modules belonging to this platform. With the Node.js platform, you do not have to install web servers such as the IIS or the Apache HTTP server. The reason is because the platform comes with a built in library which can be used as a web server. The platform has gained much popularity for use on the server side of web applications. Most companies use it for this purpose. Ryan Dahl and other developers were behind the invention of Node.js in the year 2009. It was first published in Linux in this year.

# Chapter 2:
# Environmental Setup

Before beginning to program in Node.js, you need to begin by
setting up the programming environment. You need to know
that there are several programming environments which have
been set up online for free use, so you can take advantage of
that. However, it is also good for you to know how to locally set
up the environment for this.

Local Environmental Setup

This requires you to have two software programs, which include:

- Text Editor and

- The Node.js binary installables.

# The Text Editor

This is where you will type your programs. Some of the command editors which can be used for this purpose include the Notepad in Windows, Epsilon, Brief, Vim, or the Vi and EMACS in OS X. The files having your programs will be written in these editors, and they will be said to contain the source code.

# Node.js Runtime

The source file in this file will have a Javascript code. For interpretation and execution of the code, we use the Node.js interpreter. The Node.js has distributions which come in the form of binaries, and these can be installed on the various operating systems such as Windows, Linux, OS X, and SunOS.

You need to begin by downloading the Node.js archive which is available online for free download. Make sure that you download the latest release of this.

# Installation on Linux/UNIX/ SunOS and Mac OS X

After downloading the Node.js archive corresponding to your operating system,, extract the contents of the package, and then move them to the "*/usr/local/nodejs*" directory. An example is given in the figure below:

```
$ cd /tmp
```

We are now in the "*/tmp*" directory. Let us extract the contents of the downloaded package:

```
$ wget http://nodejs.org/dist/v0.12.0/node-v0.12.0-linux-x64.tar.gz
$ tar xvfz node-v0.12.0-linux-x64.tar.gz
```

We now need to move the extracted files to the directory we have mentioned above. We should first create the directory. This can be done as follows:

```
$ mkdir -p /usr/local/nodejs
```

The directory is now ready, thus, we can move our files to it as follows:

```
$ mv node-v0.12.0-linux-x64/* /usr/local/nodejs
```

The "*bin*" folder of the above directory should then be added to the "*PATH*" variable.

# Node.js Installation on Windows

In this case, you should use the MSI file. You just have to run it as administrator, and it will execute well. If the command prompts are opened, just restart them, and the installation will take effect.

# Hello Program

We now need to verify our installation by writing and executing the *"Hello there!"* program. Open your text editor, and then create a new file. Give it the name *"test.js"*. Add the following code to your file:

**/\* Hello there! program in node.js \*/**

**console.log("Hello there!")**

Once you have written the program, just save it and then use the Node.js interpreter to execute it. Open the command prompt, and then execute the following command:

**node test.js**

After hitting the *"Return"* key, you will observe the following output:

```
Hello there!
```

This shows that the installation of the package on your machine was successful. In case you get errors, just repeat the installation process as it might have encountered some errors.

# Chapter 3:

# Creating our First Application

You need to understand the parts which make up a Node.js application. These include the following:

1.                        Importation of required modules- these are imported by use of the *"require"* directive.

2.                        The server- this receives requests from the clients, and takes the necessary action by providing a response.

3.                        Read request and return response- this is the response sent by the client to the server. The server reads the request, and then relays a response to the client.

## Loading the modules

In this case, we use the *"require"* directive. The obtained HTTP instance is turned into an http variable as shown below:

**var httpVar = require("http");**

## Creation of the server

In this, the http instance which we have created is used to call the method *"http.createServer()"* which creates an instance of the server. This is then bound to the port number 8081 by use of the *"listen"* method, which is associated to the server. Consider the example given below:

**httpVar.createServer(function (request, response) {**

**// sending a HTTP header**

**// HTTP Status: 200 : OK**

**// Content Type: text/plain**

```
response.writeHead(200, {'Content-Type':
'text/plain'});

// Sending the response body as "Hello there"

response.end('Hello there\n');

}).listen(8081);

// the message will be printed on the console

console.log('The server is running at
http://127.0.0.1:8081/');
```

We have then created our server. It will wait and listen for a request from the client via the port 8081 which is located on the local machine.

# Request and response testing

The previous steps can then be combined into a single file, and given the name *"test.js"*. The HTTP server can then be started as shown below:

```
var httpVar = require("http");

httpVar.createServer(function (request, response) {

// Sending a HTTP header

// HTTP Status: 200 : OK

// Content Type: text/plain

response.writeHead(200, {'Content-Type':
'text/plain'});

// Sending the response body as "Hello there"

response.end('Hello there\n');
```

**}).listen(8081);**

**// this message will be printed by the console**

**console.log('The server is running at http://127.0.0.1:8081/');**

Once you have written the code above, just execute the file *"test.js"*. Use the following command:

**node test.js**

You will observe the following output:

```
The server is running at http://127.0.0.1:8081/
```

Alternatively, you can open your browser and then type the following URL:

**http://127.0.0.1:8081/**

This will give the following output on the browser:

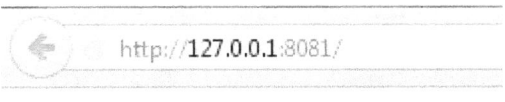

Hello there

When you get the above output, just know that your HTTP server has been set up. It is now running, and all of the incoming HTTP requests are responded to at port 8081.

# Chapter 4:

# The REPL Terminal

This stands for *"Read Eval Print Loop."* It works just the same way the Windows console or the Linux shell works. A command is entered, after which the system will provide a response to the command in an interactive mode. Node.js comes built-in with REPL. It does the following tasks:

- Read- the input from the user is read, parsed to the Javascript data structure, and then stored in the memory.

- Eval- the data structure is evaluated.

- Print- the result is printed.

- Loop- the command is looped until when the user presses "*ctrl-c*" two times.

It is of great importance when it comes to the debugging of Javascript codes and experimentation of Node.js codes.

You should learn how to start REPL. To do this, just open your console or shell, and then run the command "*node*" with no other argument. This is shown below:

```
$ node
```

The REPL command prompt will appear, and is symbolized by the "*greater than*" symbol as shown below:

```
$ node
>
```

## Running Expressions

We now need to perform some basic mathematics or calculations on the REPL terminal. Just type an addition expression on the prompt, and then hit the *"Return"* key. This is demonstrated in the figure given below:

```
> 2 + 4
6
>
```

As shown in the above figure, the result is 6, which is the answer after adding 4 to 2. Let us perform a more advanced mathematical problem:

```
> 2 + (2 * 3) - 4
4
>
```

As shown in the output, the REPL terminal knows to obey the BODMAS rule. It began by multiplying 2 and 3 where it got 6. This was added to 2 and then 4 subtracted from the result.

## Using Variables on REPL

With REPL, we can store our values in variables which can then be printed later. In case we do not use the keyword *"var,"* the value will be stored in the variable, and then printed immediately. However, if the keyword is used, the value will be stored and then not printed. To print the value of a variable, use *"console.log()"*. Consider the figure given below:

```
> x = 5
5
>
```

We used no *"var"* keyword, so the value of the variable was printed immediately after pressing the *"Return"* key. Consider the figure shown below:

```
> var y = 10
undefined
>
```

We are told that the variable is undefined. It will not be

printed immediately. Consider the figure shown below:

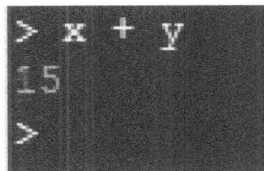

The output in the figure was the result of adding "*y*" to "*x*,"

whose values are 10 and 5 respectively. Consider the figure

shown below:

```
> console.log("Hello there!")
Hello there!
undefined
>
```

# Expressions taking multiple lines

REPL can support expressions with multiple lines, just like what happens in Javascript. Consider the example given below:

```
> var y = 0
undefined
> do {
... y++;
... console.log("The value of y:" + y);
... } while (y < 5);
```

Just write the program on the REPL as it is shown above. Once you are done, hit the "*Return*" key, and observe the output. It will be as follows:

```
The value of y:1
The value of y:2
The value of y:3
The value of y:4
The value of y:5
undefined
>
```

In the example, we have used a *"do...while"* loop. You must have noticed that the symbol "..." came automatically after the opening curly brace, that is, "{"and once you have pressed the *"Return"* key. The initial value of the variable is 1, which has formed the first value in the output. The value of the variable is also less than 5, and this is why we do not have 5 as part of the output. In each iteration, the number of the value was incremented by 1.

# Underscore variable

If you need to get the last result, use the underscore variable, that is, (_). Consider the example given below:

```
> var x = 5
undefined
> var y = 10
undefined
> x + y
15
> var result = _
undefined
> console.log(result)
15
undefined
>
```

We used the "*var*" keyword so as to declare and initialize our two variables, that is, "*x*" and "*y*." We have then printed their sum, which is 15. We have declared our third variable, that is, "*result*." Notice the use of the underscore, which means that the variable will hold our last result. We then want to use consle.log so as to print the value of the third variable, hence it gives us the value 15. Notice that 15 is the result of our last calculation.

It is good to know some of the REPL commands. These include the following:

- Ctrl + c- the current command is terminated.

- .load filename- the contents of a file are loaded to the current session of the Node REPL.

- .break- we just exit from the multiline expression.

- Ctrl + c twice- the Node REPL will be terminated.

- Tab key- shows the list of the current commands.

- .help- shows all of the available commands.

- Ctrl + d- the Node REPL will be terminated.

- Up/down keys- shows the command history with the ability to modify the previous commands.

- .save filename- the current session of the Node REPL is saved to a file.

- .clear- exits from the multiline expression.

Once you are done with the tasks you were doing on the REPL, you need to stop it. To come out of the *"Node.js"* REPL, you just have to press *"ctrl + c"* twice. This is shown below:

```
>
(^C again to quit)
>
sh-4.3$
```

# Chapter 5:
# Callbacks in Node.js

A callback is asynchronously equivalent to a function. It is called once a given task is completed. These are heavily used in Node.js. Node.js APIs are all written while taking care of callbacks, meaning that they all support this

Consider a situation where a certain function is reading a certain file. This function can return the control to an execution environment for execution of the next instruction. After completion of the I/O, the callback function will be called ,and the content of the file will be passed as the parameter. This means that there is no wait for file or block I/O. This is the reason why Node.js is highly scalable, because there is no need for it to wait for a particular function to return a result. A high number of requests can also be processed per unit time.

## Blocking Code

We need to demonstrate how a blocking code works in Node.js. Create a new text file, and give it the name *"file.txt."* Add the following content to the file:

**This book is easy to read and understand for beginners.**

**Once you read it, you will become a Node.js expert**

You can then create a Node.js file and then give it the name *"test.js."* Add the following code to it:

```
var fr = require("fs");

var d = fr.readFileSync('file.txt');

console.log(d.toString());

console.log("Program concluded");
```

**Once you have written the program, just execute it as follows:**

Open the command prompt, and then run the following command:

**node test.js**

You will observe the following output from the program:

```
This book is easy to read and understand for beginners.
Once you read it, you will become a Node.js expert
Program concluded
```

# Non-blocking code

We now need to demonstrate how a non-blocking code works in Node.js.

Create a text file named *"file.txt"* with the following code:

**This book is easy to read and understand for beginners.**

**Once you read it, you will become a Node.js expert**

Once you are done, update the file *"text.js"* to the following:

```
var fr = require("fs");

fr.readFile('file.txt', function (error, d) {

if (error) return console.error(error);

console.log(d.toString());
```

**});**

**console.log("Program concluded");**

Once you are done, open the prompt, and then execute the following command:

**node test.js**

After executing the above command, you will notice the following output:

```
Program concluded
This book is easy to read and understand for beginners.
Once you read it, you will become a Node.js expert
```

The two examples have clearly illustrated how a blocking and a non-blocking I/O works. In the first example, the program had to wait for the file reading to take place before printing the

"*Program concluded*" line. In the second example, the program does not wait for this, as it just goes ahead and prints the line "*Program concluded.*" It then resumes the task of reading the file.

To conclude, a blocking program will execute in sequence, and this makes its implementation an easy job. For the case of the non-blocking program, it is a bit difficult and tricky to implement the logic behind it. The code should be written in the same block so that the execution can be done in a sequence.

# Chapter 6:
# Event Loop

Node js is an application which is single threaded. However, with the use of callback and event, we can be in a position to create concurrency. Node js APIs are single threaded and asynchronous. This is why the framework calls the function *"async"* for the purpose of maintaining the concurrency. The thread keeps an event loop and after completion of a task, the corresponding event will be fired, which signals the event listen function for execution.

# Event Driven Programming

This framework is a bit faster, compared to other closely related frameworks. The reason behind this is the fact that the framework supports events. Once the Node server has been started, the variables are initiated, functions are declared, and then the Node waits for the event to occur.

When it comes to an event-driven application, a main loop is tasked with listening to events, and in case one of the events is detected, the callback function will be triggered. In Node.js, there are numerous events which are accessed via the *"events"* module and the *"EventEmitter"* class which can be used for the binding of event listeners and the events as shown below:

**// Importing the events module**

**var ev = require('events');**

**// Creating an eventEmitter object**

**var evEmitter = new ev.EventEmitter();**

Whenever you want to bind an event handler to an event, use the syntax given below:

**// Binding event and event handler**

**evEmitter.on('eventName', eventHandler);**

We now need to demonstrate these by use of an example.

Create a new file, and then give it the name "*test.js.*" The following is the content for the file:

**// Importing the events module**

**var ev = require('events');**

**// Creating an eventEmitter object**

```javascript
var evEmitter = new ev.EventEmitter();

// Creating an event handler

var cHandler = function connected() {

console.log('The connection was successful.');

// Firing the data_received event

evEmitter.emit('data_received');

}

// Binding the connection event to the handler

evEmitter.on('connection', cHandler);

// Binding the data_received event to the anonymous
function

evEmitter.on('data_received', function(){

console.log('The data was received succesfully.');
```

```
});

// Firing the connection event

evEmitter.emit('connection');

console.log("The program has ended.");
```

Once you have added the content to the file, just run it. Open the terminal, and then run the following command:

**node test.js**

You will observe the following output from the file:

```
The connection was successful.
The data was received succesfully.
The program has ended.
```

# The functionality of Node.js Applications

In an application created using Node js, an async function has to accept a callback as the last parameter while the callback function accepts an error as its first parameter. This can be effectively demonstrated by use of the example we gave previously. Just create a new text file and give it the name *"file.txt."* The following should be the content for the file:

**This book is easy to read and understand for beginners.**

**Once you read it, you will become a Node.js expert**

The following should be the content for the file *"test.js"*:

**var fr = require("fs");**

**fr.readFile('file.txt', function (error, d) {**

**if (error){**

```
console.log(error.stack);

return;

}

console.log(d.toString());

});

console.log("Program concluded");
```

Just run the program, and you will observe the following as the output:

```
Program concluded
This book is easy to read and understand for beginners.
Once you read it, you will become a Node.js expert
```

# Chapter 7:
# Event Emitter

Most of the Node objects have the capability to emit events. A good example is the *"net.server"* object, which will emit an event whenever a peer is connected to it. Another example is the *"fs.readStream"* which will emit an event whenever a file is opened. Objects which have the capability of emitting events are an instance of *"events.EventEmitter."*

## The "EventEmitter" class

This class is contained in the *"events"* module. To access it, we use the following syntax:

**// Importing the events module**

**var ev = require('events');**

**// Creating an evEmitter object**

**var evEmitter = new ev.EventEmitter();**

Whenever an instance of EventEmitter encounters an error, an *"error"* event will be emitted. After adding a new listener, the *"newListener"* event will be fired. After removal of a listener, the event *"removeListener"* will be fired.

We need to demonstrate how this works by use of an example. Just create a new js file and give it the name *"test.js."* The following should be the contents of the file:

**var ev = require('events');**

**var evEmitter = new ev.EventEmitter();**

**// listener #1**

**var l1 = function l1() {**

```javascript
console.log('l1 has been executed.');

}

// l #2

var l2 = function l2() {

console.log('l2 has been executed.');

}

// Binding the connection event to the l1 function

evEmitter.addListener('connection', l1);

// Binding the connection event to the l2 function

evEmitter.on('connection', l2);

var evListeners =
require('events').EventEmitter.listenerCount(evEmit
ter,'connection');
```

```javascript
console.log(evListeners + " Listner(s) are now
listening to the connection event");

// Firing the connection event

evEmitter.emit('connection');

// Removing the binding of l1 function

evEmitter.removeListener('connection', l1);

console.log("L1 will no longer listening.");

// Firing the connection event

evEmitter.emit('connection');

evListeners =
require('events').EventEmitter.listenerCount(evEmit
ter,'connection');

console.log(evListeners + " Listner(s) are listening to
the connection event");
```

**console.log("The program has ended.");**

Just open the terminal, and execute the above program. This can be done by running the following command:

**node test.js**

You will observe the following output from the program:

```
2 Listner(s) are now listening to the connection event
l1 has been executed.
l2 has been executed.
L1 will no longer listening.
l2 has been executed.
1 Listner(s) are now listening to the connection event
The program has ended.
```

# Chapter 8:
# Buffers

Pure Javascript is very friendly to Unicode, but not friendly to binary data. In Node js, we normally deal with file systems. This shows the necessity for us to handle octet streams. The Buffer class in Node provides us with instances in which we can store. This is done similarly to what happens with an array of integers.

This class is a global one, meaning that it can freely be used in an application without the need to import its module.

## Creation of Buffers

There are a variety of ways that we can construct a Node buffer. Let us discuss these ways:

1. We can create an uninitiated buffer. Consider the example given below which creates an uninitiated buffer having 5 octets:

   **var buffer = new Buffer(5);**

2. We can also create a Buffer from an array that we have. An example of this has been given below:

   **var buffer = new Buffer([50, 60, 70, 80, 90]);**

3. We can also create a Buffer from a string. An example is given below:

   **var buffer = new Buffer("Node is easy to learn", "utf-8");**

   Notice that we have used an encoded string type in the above example. Other types of encodings other than the *"utf-8"* can also be used.

# How to write to Buffers

If you need to create a method which will write to a Buffer, kindly use the syntax given below:

**buf.write(string[, offset][, length][, encoding])**

The following parameters have been used in the above syntax:

- String- this is the string which is to be written to our Buffer.

- Offset- this is the index of the Buffer where we should writing at. The default value for this is 0.

- Length- this refers to the number of bytes which are to be written. The default for this is "buffer.*default*."

- Encoding- this is the encoding which is to be used. The default for this is the "*utf8*."

Note that the method will give us a number of octets. If the number does not provide enough space, then only part of the string will be written. Consider the example given below:

**buffer = new Buffer(256);**

**length = buffer.write("Node is easy to learn");**

**console.log("Octets written are: "+ length);**

Just run the above program, and observe the output. It will be as follows:

```
Octets written are: 21
```

# How to Read from Buffers

For us to read data from the Node Buffer, we use the following syntax:

**buffer.toString([encoding][, start][, end])**

The following parameters have been used in the syntax:

- Encoding- this is the encoding to be used. The default for this is "*utf8.*"

- Start- this is the index which we should begin at. The default for this is 0.

- End- this is the index at which to stop reading. This is set to the complete string by default.

What the method does is that it decodes and then returns a string from the buffer data which is encoded in the data encoding which you had specified. Consider the example given below:

```
buffer = new Buffer(26);

for (var j = 0 ; j < 26 ; j++) {

buffer[j] = j + 97;

}

console.log( buffer.toString('ascii'));      // This will
output: abcdefghijklmnopqrstuvwxyz

console.log( buffer.toString('ascii',0,10));   // This
will output: abcdefghij

console.log( buffer.toString('utf8',0,10));   // This
will output: abcdefghij

console.log( buffer.toString(undefined,0,10)); // the
default encoding style //will be  'utf8', and then
output abcdefghij
```

Just write the program just as it is, and then run it. You will observe the following output:

```
abcdefghijklmnopqrstuvwxyz
abcdefghij
abcdefghij
abcdefghij
```

## Conversion of Buffer to JSON

For you to convert a Node Buffer to a JSON object, use the syntax given below:

**buffer.toJSON()**

In this case, the returned value will be a Buffer instance represented as a JSON object. Consider the example given below:

**var buffer = new Buffer('Node js is easy to learn');**

**var json = buffer.toJSON(buffer);**

**console.log(json);**

Just write the program, and then run it. You will observe the following output:

[ 78, 111, 100, 101, 32, 106,
115,32,105,115,32,101,97,115,121,32,116,111,32,108,101,97,114,1
10 ]

# Concatenation of Buffers

Node Buffers can be concatenated into a single Node Buffer.

This takes the following syntax:

**Buffer.concat(list[, totalLength])**

The following parameters have been used:

- List- this is the list of elements to be concatenated.

- totalLength- this is the length of the Buffer after concatenation.

The return value of this method is an instance of the Buffer.

Consider the example given below:

**var b1 = new Buffer('Node js');**

**var b2 = new Buffer('is easy to learn');**

**var b3 = Buffer.concat([b1,b2]);**

```
console.log("The content of b3 is: " + b3.toString());
```

Just write and then run the above program. You will observe
the following as the output:

```
The content of b3 is: Node jsis easy to learn
```

# Comparison of Buffers

It is possible for us to compare two Buffer Nodes. This takes the following syntax:

**buffer.compare(otherBuffer);**

The following parameter has been used:

- otherBuffer- this is the Buffer which is to be compared with the *"buffer."*

The return value for this method is a number which indicates whether it comes after or before or the buffers are the same in sorted order. Consider the example given below:

**var b1 = new Buffer('WXY');**

**var b2 = new Buffer('WXYZ');**

**var answer = b1.compare(b2);**

**if(answer < 0) {**

```javascript
console.log(b1 +" comes before " + b2);

}else if(answer == 0){

console.log(b1 +" is the same as " + b2);

}else {

console.log(b1 +" comes after " + b2);

}
```

Just write the above program, and then run it. You will observe the following output:

```
WXY comes before WXYZ
```

# Copying the Buffer

If you need to copy a node buffer, then use the following syntax:

**buffer.copy(targetBuffer[, targetStart][, sourceStart][, sourceEnd])**

The following parameters have been used in the above syntax:

- targetBuffer- this is the Buffer object where the Buffer is to be copied.

- targetStart- this is an optional number whose default value is 0.

- sourceStart- this is an optional number whose default value is 0.

- sourceEnd- this is an optional number whose default value is buffer.length.

This method has no return value. Data will be copied from the source Buffer to the target Buffer even when the memories of the two overlap. If the targetStart and the sourceTarget are not defined, then their value will be defaulted to 0.

Consider the example given below:

**var b1 = new Buffer('XYZ');**

**//copying a buffer**

**var b2 = new Buffer(3);**

**b1.copy(b2);**

**console.log("The content of b2 is: " + b2.toString());**

After executing the above program, the following will be the output:

```
The content of b2 is: XYZ
```

What has happened is that the content of buffer b1 has been copied to the buffer b2. They now have the same contents.

# Slice Buffer

In this case, our intention is to get a sub-buffer of a particular node buffer. It takes the following syntax:

**buffer.slice([start][, end])**

The following parameters have been used in the above syntax:

- Start- this is an optional number whose default value is 0.

- End- this is an optional number whose default value is *"buffer.length."*

The return value of this method is a new buffer referencing the same memory location as the old buffer, but it will be cropped and offset depending on the values of *"start"* and *"end."* Note if the indexes are negative, then they will begin at the end of the buffer.

Consider the example given below:

**var b1 = new Buffer('Node js is easy');**

**//slicing a buffer**

**var b2 = b1.slice(0,9);**

**console.log("The content of b2 is: " + b2.toString());**

Just execute the above program, and you will get the following output:

```
The content of b2 is: Node js i
```

As shown in the above figure, we only have part of the sentence which we had specified. The reason is because we have sliced it.

# Buffer Length

This method is used to get the size of a node buffer in bytes. It takes the following syntax:

**buffer.length;**

Note that the return value of this method is the size of the buffer in terms of bytes. Consider the program given below:

**var b = new Buffer('Node js');**

**//length of the buffer**

**console.log("The length of the buffer is: " + b.length);**

Just run the above program, and you will observe the following output:

```
The length of the buffer is: 7
```

The program has given us the length of the buffer in terms of bytes, meaning that our string is 7 bytes long.

# Chapter 9:
# Streams in Node.js

The purpose of streams is to assist programmers to continuously write data to a destination or read the data from a particular source. Node js supports four types of streams. These include the following:

- Readable- these are streams which are used for a read operation.

- Writable- these are streams which can be used for writing operation.

- Duplex- these are the kind of streams which support both read and write operations.

- Transform- these are streams in which the calculation of the output is done based on the input.

A stream is just an instance of an *"EventEmmiter"* and it throws several events at different instances of time.

# How to Read from Stream

We now need to demonstrate how reading from a stream can be done. Just create a new text file, and give it the name *"file.txt."*. Add the following content to the file:

**Node js is easy to learn. You only need to be interested.**

**This book will help to make you an expert in Node js.**

After creating the text file and adding the content to it, create a js file and give it the name *"test.js."* The following should be its content:

**var fr = require("fs");**

**var d = '';**

**// Creating a readable stream**

**var rStream = fr.createReadStream('file.txt');**

```javascript
// Setting the encoding to utf8.

rStream.setEncoding('UTF8');

// Handling stream events --> data, end, and error

rStream.on('d', function(chunk) {

d += chunk;

});



rStream.on('end',function(){

console.log(d);

});

rStream.on('error', function(error){

console.log(error.stack);
```

```
});
```

```
console.log("Program terminated");
```

Once you have added the content to the file, just execute it.

You will observe the following output:

```
Program terminated
Node js is easy to learn. You only need to be interested.
This book will help to make you an expert in Node js.
```

# Writing to Stream

We want to demonstrate how writing to a stream can be done.
Just create a js file with the name *"test.js."* The following
should be the content of the file:

**var fr = require("fs");**

**var d = 'Node js is very easy';**

**// Creating a writable stream**

**var wStream = fr.createWriteStream('file.txt');**

**// Writing the data to the stream with encoding as
utf8**

**wStream.write(d,'UTF8');**

**// Marking the end of the file**

```
wStream.end();

// Handling stream events --> finish, and error

wStream.on('finish', function() {

console.log("Writing has completed.");

});

wStream.on('error', function(error){

console.log(error.stack);

});

console.log("Program terminated");
```

Once you have written the program, just run it and you will
observe the following output:

```
Program terminated
Writing has completed.
```

In the current directory, look for the file with the anme "*file.txt.*" This has been created by the program. Open it, and you will find it having the text "*Node js is very easy.*" This is the text which we specifed that it should be written to our file.

## Piping of streams

In stream piping, we make the output from one stream to be the input for the other stream. We used it for getting data from a stream which is then passed to the next stream. Note that we have no limit to piping operation.

In the example to be given, we will read from a particular file, and then write what we have read to another file.

Just create a js file and give it the name *"test.js."* Add the following code to it:

**var fr = require("fs");**

**// Creating a readable stream**

**var rStream = fr.createReadStream('file.txt');**

**// Create a writable stream**

**var wStream = fr.createWriteStream('out.txt');**

**// Piping read and the write operations**

**// reading file.txt and writing the data to out.txt**

**rStream.pipe(wStream);**

**console.log("Program terminated");**

Just run the above program, and then observe the output. It will be as follows:

```
Program terminated
```

In your current, just find the file "*out.txt.*" This file has been created by the above program. Open it, and you will find it having the following text:

**Node js is easy to learn. You only need to be interested.**

**This book will help to make you an expert in Node js.**

# Stream Chaining

In this operation, the output from one stream is connected to another stream in which we will form a chain of stream operations. The operation is used together with the piping operation. In the example to be given, we use these two operations so as to compress a particular file and then decompress it.

Create a js file and give it the name *"test.js."* The following should be the code for the file:

```
var fr = require("fs");

var complib = require('zlib');

// Compressing the file file.txt to in.txt.gz

fr.createReadStream('file.txt')

.pipe(complib.createGzip())
```

**.pipe(fr.createWriteStream('file.txt.gz'));**

**console.log("File has been Compressed.");**

Just write and then run the above program. You will get the following output:

```
File has been Compressed.
```

Remember that the following is the content for the file *"file.txt"*:

**Node js is easy to learn. You only need to be interested.**

**This book will help to make you an expert in Node js.**

In the same directory, that is, the current directory, you will find a compressed file with the name *"file.txt.gz."* This is an indication that our program worked effectively. We now need to reverse the process, that is, decompress the file. The following code can be used for this purpose:

```
var fr = require("fs");

var complib = require('zlib');

// Decompressing the file file.txt.gz to file.txt

fr.createReadStream('file.txt.gz')

.pipe(complib.createGunzip())

.pipe(fr.createWriteStream('file.txt'));

console.log("File has been Decompressed.");
```

Just write the above code, and then execute it. You will observe the following output:

```
File has been Decompressed.
```

# Chapter 10:

# The File System

The file I/O is implemented in Node by use of simple wrappers around the standard POSIX functions. The module for the file system in Node can be implemented by use of the following syntax:

**var fr = require("fs")**

## Synchronous vs Asynchronous

Fs methods has both the synchronous and asynchronous forms. With the asynchronous methods, the last parameter is taken as the completion function callback, while the first parameter of the callback function becomes the error. With the asynchronous method, there is no blocking of the execution of a program as compared to what happens with the synchronous methods.

Let us demonstrate this by use of a program.

Create a new text file, and give it the name "*file.txt.*" Add the following content to the file:

**Node js is easy to learn. You only need to be interested.**

**This book will help to make you an expert in Node js.**

Once you have done the above, create a js file and give it the name "*test.js.*" The following code should be added to the file:

**var fr = require("fs");**

**// Asynchronous reading**

**fr.readFile('file.txt', function (error, d) {**

**if (error) {**

```
    return console.error(error);

}

console.log("Asynchronous reading: " + d.toString());

});

// Synchronous reading

var d = fr.readFileSync('file.txt');

console.log("Synchronous reading: " + d.toString());

console.log("Program terminated");
```

Just write the above program, and then execute it. You will notice the following output:

```
Synchronous reading is: Node js is easy to learn. You only need to be interested.
This book will help to make you an expert in Node js.
Program terminated
Asynchronous reading is: Node js is easy to learn. You only need to be interested.
This book will help to make you an expert in Node js.
```

The program has read the contents of the file both synchronously and asynchronously. The difference between the outputs from the two kinds of read is the concept of blocking. This was discussed previously.

# Opening File

If you need to open a file in an asynchronous mode, then use the following syntax:

**fr.open(path, flags[, mode], callback)**

The following parameters have been used in the above syntax:

- Path- this is the string which contains the name of the file and its path.

- Flags-this tells more of the behavior of the file which is to be opened.

- Mode- this is used for setting the mode of the file if it has been created.

- Callback- this is the callback function which is to be used for getting our two arguments, that is, d and error.

# Flags

Flags can be used for both the read and write operations.

Let us demonstrate how these are used by use of an example.

Create a new js file, and give it the name *"test.js."* Add the following code to the file:

```
var fr = require("fs");

// Asynchronous opening of the file

console.log("Beginning to open the file!");

fr.open('file.txt', 'r+', function(error, f) {

if (error) {

return console.error(error);

}
```

**console.log("The file was opened successfully!");**

**});**

Just write the above program, and then execute it. To execute it, just open the terminal and then type the command:

**node test.js**

You will observe the following output:

```
Beginning to open the file!
The file was opened successfully!
```

The output shows that the file was opened successfully.

# Getting File Information

If you need to get information about a certain file, use the following syntax:

**fr.stat(path, callback)**

The following parameters have been used:

Path- this is the string specifying the file name and its path.

Callback- this is the function which is to get information about our two arguments. Let us demonstrate this by use of an example.

Create a js file and give it the name *"test.js."* Add the following code to the file:

**var fr = require("fs");**

**console.log("Beginning to get information about the file!");**

```
fr.stat('file.txt', function (error, statistics) {

if (error) {

return console.error(error);

}

console.log(statistics);

console.log("The file info was obtained
successfully!");

// Checking the file type

console.log("isFile ? " + statistics.isFile());

console.log("isDirectory ? " +
statistics.isDirectory());

});
```

Just run the program once you have written it, and then observe the output that you get. It will made up of the details of the file. The first part of the output will be as follows:

```
Beginning to get information about the file!
```

The rest will be a description of the file.

# Writing to File

If you need to write to a file, use the following syntax:

**fr.writeFile(filename, data[, options], callback)**

If the file is found to exist, then it will be overwritten. This is why this method is not recommended when writing to an already existing file. The following parameters have been used in the above syntax:

- Path- this is the string which specifies the name of the file and its path.

- Data- this is the string or the Buffer which is to be written to the file.

- Option- this is an object which will hold the flag, encoding, and the mode. These have their own default values.

- Callback- this is the method to get the parameter and then return an error if it occurs.

Let us demonstrate this by use of a program.

Open the text editor, and then create a js file named *"test.js."* Add the following content to the file:

```
var fr = require("fs");

console.log("Beginning to write to a file which is existing");

fr.writeFile('file.txt', 'Node js is easy to learn!', function(error) {

if (error) {

return console.error(error);

}

console.log("The data was written successfully!");
```

```
console.log("Let us read the newly written data");

fr.readFile('file.txt', function (error, d) {

if (error) {

return console.error(error);

}

console.log("Asynchronous reading gives: " +
d.toString());

});

});
```

Once you have written the above program, just save and then run it. You will observe the following output:

```
Beginning to write to a file which is existing
The data was written successfully!
Let us read the newly written data
Asynchronous reading gives: Node js is easy to learn!
```

# Reading File

If you need to read from a particular file, then use the following syntax:

**fr.read(fd, buffer, offset, length, position, callback)**

The file descriptor is used for the purpose of reading the file. However, if you need to use the exact file name for this purpose, then there are other methods which can be used for this purpose. The following are the parameters which have been used:

- Fd- this is the file descriptor, and it is returned by the "*fr.open()*" method.

- Buffer- this is the buffer in to which we will write the data.

- Offset- this is where writing in the buffer should begin.

- Length- this is an integer which specifies the number of bytes which are to be read from the file.

- Position- this is an integer which specifies where reading should be started. If null, the reading will be begun from the current position.

- Callback- this method will get our three arguments.

We want to demonstrate this by use of a program. Just create a js file named *"test.js"* and then add the following code to it:

**var fr = require("fs");**

**var buffer = new Buffer(1024);**

**console.log("Starting to open an already existing file");**

**fs.open('input.txt', 'r+', function(error, fd) {**

**if (error) {**

```javascript
        return console.error(error);

}

console.log("The file was opened successfully!");

console.log("Begining to read the file");

fr.read(fd, buf, 0, buf.length, 0, function(error, bytes){

if (error){

console.log(error);

}

console.log(bytes + " bytes were read");

// Printing only the read bytes to avoid junk.

if(bytes > 0){

console.log(buffer.slice(0, bytes).toString());
```

```
}
```

```
});
```

```
});
```

Just write the program, and then execute it. You will observe the following output:

```
Starting to open an already existing file
The file was opened successfully!
Begining to read the file
94 bytes were read
Node js is easy to learn. You only need to be interested.
This book will help to make you an expert in Node js.
```

As shown in the above figure, the file which we specified has been read successfully.

## Closing File

Once you have opened a file, and you are done with whatever you were doing, you may need to close it. The method takes the following syntax:

**fr.close(fd, callback)**

The following parameters have been used in the above syntax:

- Fd- this is the file descriptor which the method *"fr.open()"* returns.

- Callback- this is the callback function which will get no argument.

Let us use an example so as to demonstrate this. Create a js file, and give it the name *"test.js."* Add the following code to the file:

**var fr = require("fs");**

**var buffer = new Buffer(1024);**

```
console.log("Starting to open an already existing
file");

fr.open('file.txt', 'r+', function(error, fd) {

if (error) {

return console.error(error);

}

console.log("The file was opened successfully!");

console.log("Starting to read the file");

fr.read(fd, buf, o, buf.length, o, function(error,
bytes){

if (error){

console.log(error);

}
```

```javascript
// Print only read bytes to avoid junk.

if(bytes > 0){

console.log(buffer.slice(0, bytes).toString());

}

// Closing the opened file.

fr.close(fd, function(error){

if (error){

console.log(error);

}

console.log("The file was closed successfully.");

});

});
```

**});**

Just write the program as it is shown above, and then run it.

You will observe the following output:

```
Starting to open an already existing file
The file was opened successfully!
Starting to read the file
Node js is easy to learn. You only need to be interested.
This book will help to make you an expert in Node js.
The file was closed successfully.
```

From the output given, it is very clear that the file was opened, read, and then closed. The program worked effectively.

# Truncating File

It is possible for us to truncate a file which has been opened.

This takes the following syntax:

**fr.ftruncate(fd, len, callback)**

The following parameters have been used:

- Fd- this is the file which the method *"fr.open()"* returns.

- Len- this defines the length after which we will have to truncate our file.

- Callback- this is the callback, and it will get no argument.

Let us use a program to demonstrate how this can be done. Create a js file, and give it the name *"test.js."* Add the following code to the file:

**var fr = require("fs");**

```javascript
var buffer = new Buffer(1024);

console.log("Starting to open an already existing file");

fr.open('file.txt', 'r+', function(error, fd) {

if (error) {

return console.error(error);

}

console.log("The file was opened successfully!");

console.log("Starting to truncate the file after 12 bytes");

// Truncating the opened file.

fr.ftruncate(fd, 10, function(error){

if (error){
```

```javascript
        console.log(error);

    }

    console.log("The file was truncated successfully.");

    console.log("Starting to read the file");

    fr.read(fd, buffer, 0, buffer.length, 0, function(error,
    bytes){

        if (error){

            console.log(error);

        }

        // Printing only the read bytes to avoid junk.

        if(bytes > 0){

            console.log(buffer.slice(0, bytes).toString());

        }
```

```
// Closing the opened file.

fr.close(fd, function(error){

if (error){

console.log(error);

}

console.log("The file was closed successfully.");

});

});

});

});
```

Just write the program, and then run it. You will observe the following output:

```
Starting to open an already existing file
The file was opened successfully!
Starting to truncate the file after 12 bytes
The file was truncated successfully.
Starting to read the file
Node j
The file was closed successfully.
```

As shown in the above figure, only the part shown forms the first 12 bytes of the file. This is why they form the output.

# Deleting File

The method for deleting a file takes the following syntax:

**fr.unlink(path, callback)**

The following parameters have been used:

- Path- this is the name of the file to be deleted and its path.

- Callback- this forms our callback function, taking no argument.

Let us demonstrate this by use of an example. Create a new js file named *"test.js."* Add the following code to the file:

**var fr = require("fs");**

**console.log("Starting to delete the existing file");**

**fr.unlink('file.txt', function(error) {**

```
if (error) {

return console.error(error);

}

console.log("The file was deleted successfully!");

});
```

Just write the program, and then run it. You will observe the following output:

```
Starting to delete the existing file
The file was deleted successfully!
```

The output shows that we successfully deleted the file.

# Conclusion

It can be concluded that Node js is a very important platform as far as programming is concerned. The framework uses an I/O mechanism which is event-driven and lightweight. This makes it very easy and fast to learn, even for beginners. The platform was developed to help developers in the development of network applications which are easy to scale. It is also open-source, meaning that it is available for a free download and use. It is suitable for the development of applications which are to operate in real-time and are data intensive.

The platform exhibits a high degree of flexibility. The reason is because it is supported in nearly all of the operating systems which are commonly used by developers. It comes with an in-built library which acts as a web server itself. This means that the platform can be used a web server itself without the need for us to install web servers such as the HTTP web server.

Before starting to program in Node js, you need to begin by setting up your development environment. A text editor is needed ,and this will provide you with the environment for writing your code. Depending on the operating system that you are using, chose the best text editor. For Windows users, Notepad is good, For Mac OS X users, EMACS editor is good. If you are using Linux, then the Vim editor is good. You then need to download the installables for Node js. You can then install these after extracting them, since they will be downloaded in a zipped format.

The installation steps for Linux and Mac OS X are almost the same. The command line should be used for extraction of the package. Installing this in Windows is easy, as you just have to run the package. A Node js application has numerous parts including the server. This book has discussed all of these. My hope is that you now have a good understanding of Node js.